# Copyright © 2021 by GM Publishers

The copyright of this book is registered by GM Publishers. None can publish this book or part of it without the permission of the publisher or author. No part of this publication may be reproduced, stored in a retrieval system, or transmitted in any form or by any means, electronic, mechanical, photocopying, reading, or otherwise, without the permission of the publisher or author. If anyone copy, publish, print and plagiarized the book will be illegal offence in the eye of law and will be punished.

All rights are reserved. Published by GM Publishers

# Climate Change
## The Roles of Govt., Industries, NGOs, Political Parties, Media & Public

**Author:** G. Hossain

**Co-Authors:** Md. Fazle Mubin

**Designer:** G. Hossain

**Publications Format:** Kindle E-Book Format, Paperback Format

**Edition No:** First Edition, July, 2021

**Publication From:** Dhaka, Bangladesh

**Version:** International Version

**Published by:** GM Publishers

**ISBN:** 9798547022685

**Email address:** gmpublishers04@gmail.com

# Table of Contents

1.0 Introduction..................................................4

1.1 Background of Climate Change and Relationship to the Organizational Policy Development...................8

1.2 Climate Change Induced Natural Disasters Around the Globe......................................................9

2.0 Key Issues, Positions and Actors Influencing Public Opinion Formation Regarding the Climate Change Issue..........................................................12

2.1 Key Issues Regarding Climate Change..............12

2.2 Key Positions and Factors..............................14

2.2.1 NGOs....................................................14

2.2.2 Political Party & Politicians........................16

2.2.3 Key Media Source & Influential Media Commentators................................................16

2.2.4 Industrial Body.........................................18

3.0 Discussion and Analysis of Its Significance in Relation to Government Action...........................20

4.0 Forecast of Key Trends of the Climate Change and Its Relationship to the Organizational Policy Development...................................................20

5.0 Climate Change & Its Adverse Effects on Hobert City..............................................................21

5.1 Natural Impact............................................21

5.1.1 Changes in Perception...............................21

5.1.2 Sea Level Rise.........................................23

5.2 Social Impact……………………………………..24

5.2.1 Water Supplies……………………………….....24

5.2.2 Food Supplies…………………………………..25

5.2.3 Refugee Movement……………………………..25

6.0 Climate Change Adaptation Plan for Hobert City……………………………………………………..26

6.1 Technological/ Structural…………………………26

6.2 Food & Water Management………………………27

6.3 Policy & Governance……………………………..28

7.0 Conclusion………………………………………..29

8.0 References………………………………………..30

About the Authors……………………………………..35

   G. Hossain……………………………………….35
   Mohammed Fazle Mubin………………………….36
Also by G. Hossain & GM Publishers………………..37

## 1.0 Introduction

Since the birth of mankind, the world has changed quite a bit because we have used the resources of nature in so many ways. The world has seen so much change in its climate for that reason but things have turned much worse over the couple of centuries. Since the introduction of the Industrial Revolution the world has taken the turn from being agricultural central to industrial central (Slezak, 2019). That means there was rise of immense number of industries from that point onwards.

This urbanization and industrialization gave rise to much big and successful company which has taken the top position like the company we have discussed in this book. The company in focus is a beverage company with a huge industrial reach (Cha & Lee, 2017).

That's why the company needs very large amount of supply and resources from the nature. This need and using up of the resources of the nature definitely has a negative effect on the nature but also taking the industrial waste of the company into account then the damage really stacks up. Climate change has been a big concern for humans for a while now. This change was mainly accelerated because of the excessive rise of industries over the centuries. The different needs that these industries require starting from water to trees have been affecting negatively to the environment (Droege, 2012). The excretion of industrial waste into the environment also stacks on the damage as well. The different gases that these industries give out to the environment can cause a permanent damage to the climate.

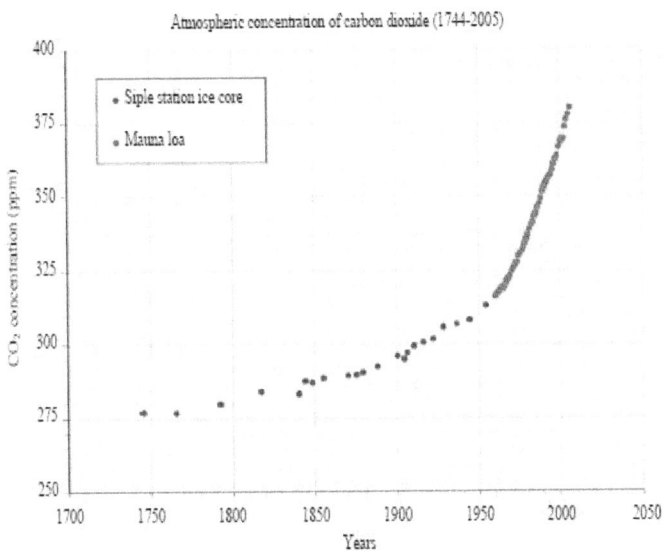

**Fig 1: Atmospheric Concentration of $CO_2$**

These gases go out into the environment making the climate hotter and trapping the heat into the atmosphere. This heat also causes the ice to melt and flooded the low-level areas of the world (Slezak, 2019). Over the last two centuries from 1880 to now and onwards it was seen that the total temperature increase alone in these years was 0.8 degrees. This might look like a small number but the effects and result of this increase of temperature is massive when it comes into play with the climate change.

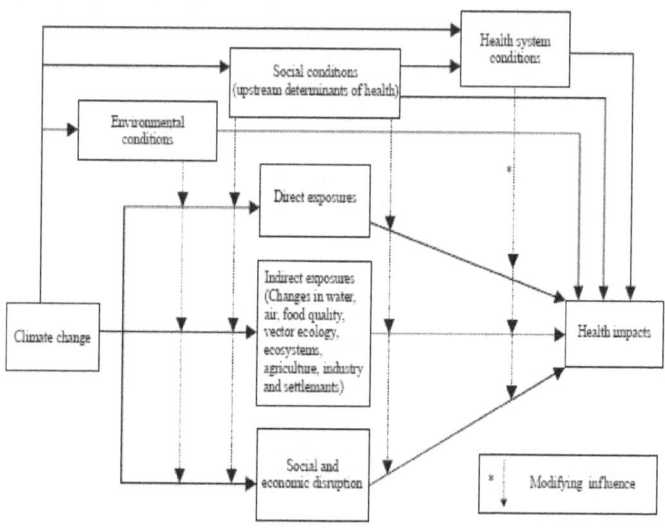

**Fig 2: Climate Change and Its Impact on Environment and Society**

Climate change is directly or indirectly impacting our regular lives and the climate condition of the different regions is becoming violent day by day. It makes the natural disasters and hazards more extreme than the past. The emission of Green House Gases (GHGs) is increasing the temperature of the atmosphere and polluted it as well.

For this reason, the temperature of earth surface is gradually increasing (Crowley, 2017).

Now we're going to discuss the adverse effects of climate change in Australia, especially Hobart City. Most of the urbanizations in Australia are seen at the coastal areas of the region, for that reason most of the cities of Australia such as Sydney, Melbourne, NSW, Hobart etc. are mostly vulnerable to climate change (Birch, 2016).

Within all of these, Tasmania's capital Hobart is becoming more vulnerable to climate change as the 1% rise of sea level can inundate 45% of the city within an hour or two. Additionally, climate change can negatively impact on the natural life, community life, ecosystem and natural resources as the biodiversity can be greatly affected due to the adverse effect of the change.

The research also tries to identify the impact of climate change in Australia's Tasmania island capital Hobart. The natural, social and economic impacts of climate change on this city and the entire region are also discussed. The research also analyzes whether the govt. has taken any action or climate change adaptation strategy to cope with this major issue.

## 1.1 Background of Climate Change and Relationship to the Organizational Policy Development

Climate change is the thing which has started to take effect now more than ever which has also impacted everything on earth. The impact of climate change on the company is also very noticeable as the organizational policy has to be changed because of climate change (Slezak, 2019). This organizational policy had to be changed because of many factors that have affected the company.

We have selected a company in this research to show how a company could play a negative role in the climate change. The selected company is mainly based on beverage as its key product, so it uses a lot of water and the waste products from the industries are also let out in the environment which can be harmful for the climate. Moreover, the different gases which are emitted out in the nature can be very harmful to the environment (Cha & Lee, 2017).

That is why there should be some changes integrated in the organizational policy of the company for all these factors working together. This also creates a clear relation of the company with the climate change and the harmful effects it may have on the environment as well.

## 1.2 Climate Change Induced Natural Disasters Around the Globe

Climate change is creating more devastating natural disasters around the globe. Last few years, we have been witnessed several devastating natural disasters in South Asia, South-East Asia, East Africa, Europe, North and Central America. Climate induced disasters has been tripled in the last 30 years. Global sea level rise is now 2.5 times faster than before.

More than 20 million people lost their livelihoods, shelter and migrated to different places due to the climate change and climate change related natural disasters. The United Nations Environment Programme (UNEP) estimates that adapting to climate change and coping with damages will cost the developing countries around $140-300 billion per year by 2030.

The climate change induced natural disasters are discussed below:

**Cyclone:** The likelihood of the cyclone is increasing day by day across the different parts of the world. Recently, South Asian countries such as Bangladesh, India, Sri Lanka etc. countries have faced several devastating cyclones. Meanwhile, Zimbabwe, Malawi and Mozambique etc. African countries have faced devasting cyclones like Cyclone Idai and Kenneth (Oxfam, 2021). These devasting cyclones destroyed the lives of thousands of people and damaged many communities as well.

**Drought:** The likelihood of the drought has also been increased than before. Higher sea temperatures increase the likelihood of the drought, different parts of the world are affected by the prolonged drought in the recent years. The East African regions have faced several devasting droughts in 2011, 2017 and 2019 which have repeatedly wiped-out crops and livestock (Oxfam, 2021).

**Flash Flood:** The likelihood of the devastating flash flood has been increased across the South, South-east Asia and Europe due to heavy rainfall during the rainy seasons. Recently, EU countries such as Germany, Belgium and The Netherlands several communities have been the victims of the deadly flash flood which causes the lose of hundreds of lives and destroys the livelihoods in these regions.

The researchers of NASA said, these types of deadly flash floods are caused by intense and frequent downpours in many parts of the world (NASA, 2021). These intense and frequent downpours occur due to the global warming.

**Flood:** The frequency, intensity and likelihood of the prolonged flood have been increased in the South Asian regions like Bangladesh, India and Nepal. This deadly flood causes landslide in these regions which takes the lives of the affected people.

Last two years, heavy monsoon rains and intense flooding destroyed the lives of thousands of people and devasted their communities and livelihoods as well (NASA, 2021).

**Wildfire:** The likelihood of wildfire is also increased than ever before due to the global warming and decrease of annual rainfall in the particular parts of the world. Recently, Amazon rainforest, California, Turkey's southern coast and several states of Australia face the devastation of bushfire and wildfire. Millions of hectors of lands have been burnt out in this devasting wildfire and lots of people and animals lost their lives as well.

**Dry Corridor:** In Central America, the frequency and intensity has been increased dramatically than ever before. Central American countries called El Salvador, Nicaragua Guatemala and Honduras are now facing 6-months long dry seasons which destroy their crops and insecure their farming as well (Oxfam, 2021).

Humanitarian crisis and shortage of food are inevitable in these regions as the crop cultivation is greatly hampered due to this dry corridor

## 2.0 Key Issues, Positions and Actors Influencing Public Opinion Formation Regarding the Climate Change Issue

As it was mentioned before climate change has an effect on everything at this point so there is also a clear relation of climate change with the features of the company and how it functions as well. There are many issues that may arise when relating climate change with the company.

These issues may have negative impact on the environment for which there may be public concern regarding this effect by the company on the environment. The different issues factors and other aspects regarding the company and climate change is given below which will help to satisfy the need of the information of this report.

## 2.1 Key Issues Regarding Climate Change

There are many issues that can be noted which impact on climate change by the company. Some of these issues can be very harmful and have to be negated while the others maybe trivial problems concerning the climate change. These issues are given below:

- The usage of water by a beverage company is huge. Because a beverage company usually needs a huge amount of pure and fresh water sources to manufacture their product.

- The company takes a lot of water from different water bodies for which there are water related issues that occur. The issues like water pollution, rivers or water bodies drying out or spreading of different disease may even occur (Droege, 2012).
- Any industry when starts on production create a lot of waste products, these waste products are often dumped in the nature.
- These waste products contain very harmful chemicals which can cause many diseases and natural issues as well. The living creatures are also affected by this which makes it a serious concern.
- Another thing that is very common or any industry is the emission of harmful gases.
- These gases go out in the environment and cause pollution also when these gases are inhaled by living creatures it may cause many diseases as well. The gases which go out like $CO_2$, CO, CFC etc. corrodes the ozone layer as well. That is why it is also a major concern (Lemos & Rood, 2010).
- The people living near industries may face over heat in temperature because of the heat and gases

coming out of these industries which can be taken as a concern as well (Droege, 2012).

- The plastic bottles are not bio-degradable which causes environment pollution,

## 2.2 Key Positions and Factors

The company has many factors which are working against the environment and pollution the climate as well. Some of these factors are given below:

- The increase heat coming out of the industries is a very harmful factor.
- The fumes of the industries affect the climate in an adverse way.
- The waste products contain chemicals which is a very concerning factor.
- The people working in the industry may face respiratory issues (Slezak, 2019).
- There can be many diseases spreading if the waste materials are not cleared.
- The plastic bottles manufactured are harming the environment as they are not bio-degradable.

### 2.2.1 NGOs

NGOs have a big factor to play in the preserving of nature and safeguarding of climate:

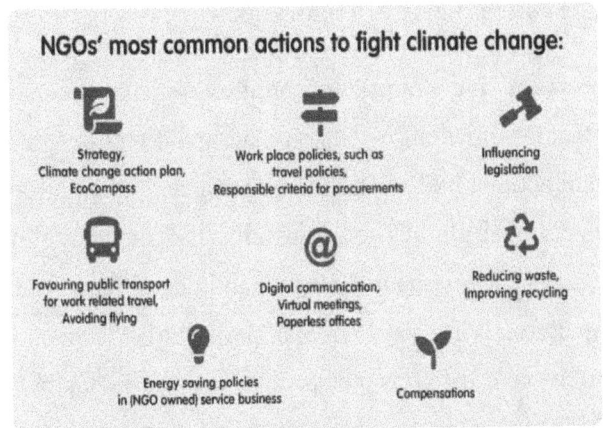

**Fig 3: NGO's Most Common Actions to Fight Climate Change**

There is also a connection of NGOs with the company as well because these NGOs pressure the company to take measures to secure the climate and preventing it from any harm by such companies.

There are many NGOs working for this sake such as BRAC, Climate Action Network (CAN), Greenpeace, World Wildlife Fund (WWF) and Oxfam etc. These NGOs and private organizations do a lot for the sake of the environment and also have impact on the activity of company like this (Lemos & Rood, 2010). These actions make these types of companies to change their policy and activities as well.

## 2.2.2 Political Party & Politicians

There is a huge impact of politics on climate change. Politics is something which can aid and direct people into certain actions like saving the environment as an instance. The political parties of Australia have made climate change as a key issue and made it an integral factor in their campaigns (WTO, 2019). The parties of Morrison and Shorten have different perspective on this factor but the politicians and parties have started acting on this issue (Cave, 2019).

The students and organizations have started to take actions to stop this manmade disaster. The companies and industries which are harming the climate are being forced to take necessary steps to protect the environment.

## 2.2.3 Key Media Source & Influential Media Commentators

There have been many issues being presented regularly on the media websites and pages which are related to climate change (Lemos & Rood, 2010). The media are doing their job to let people know about these ongoing issues and how it is affecting the climate. The media of Australia, USA, EU and UK have also gone into this issue to give the pubic a clear idea on how things are going and how the environment is being harmed as well (Cave, 2019).

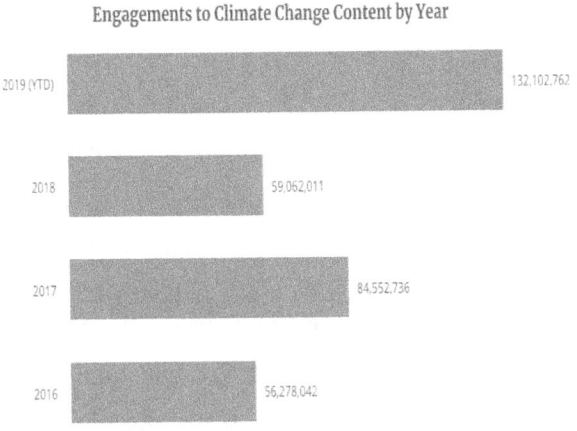

## Fig 4: Social Media Based Climate Change Content Engagements are Increasing in Every Year

Social Media can play vital role to influence the public on climate change and its adverse effects on environment through publishing/ posting climate change related contents.

Over the years, social media-based climate change content engagements are increasing. People are like to know on climate change and its negative effects by reviewing the climate change contents on Facebook, Twitter, Instagram, Youtube etc. social media networks.

In Australia, Paul Barry the host of the program called 'Media Watch' at Australian Broadcasting Corporation (ABC) showed the weaknesses of IPCC (Intergovernmental Panel on Climate Chang) policy development in his show on regular basis (ANU, 2019). Another popular editor and columnist Graham Lloyd provided his findings on IPCC's policy regarding climate change action plan in his Australian's story by Environment article (ANU, 2019).

Meanwhile, the daily telegraph columnist Bob Carter's tried to relate the climate change related policy and its interconnection with the business organizations policy formulation very effectively in his several columns and articles (ANU, 2019). These people and medias are mostly influencing regarding the climate change and organizational policy formulation issues.

### 2.2.4 Industrial Body

Industries can be considered as the major sources of harming environment and the climate. The industries throw their wastes and emit gases those are harmful for the climate. Many powerful industrial bodies can be taken as an origin for causing these issues but some industrial bodies have also risen to make sure that these harms are not caused anymore (WTO, 2019).

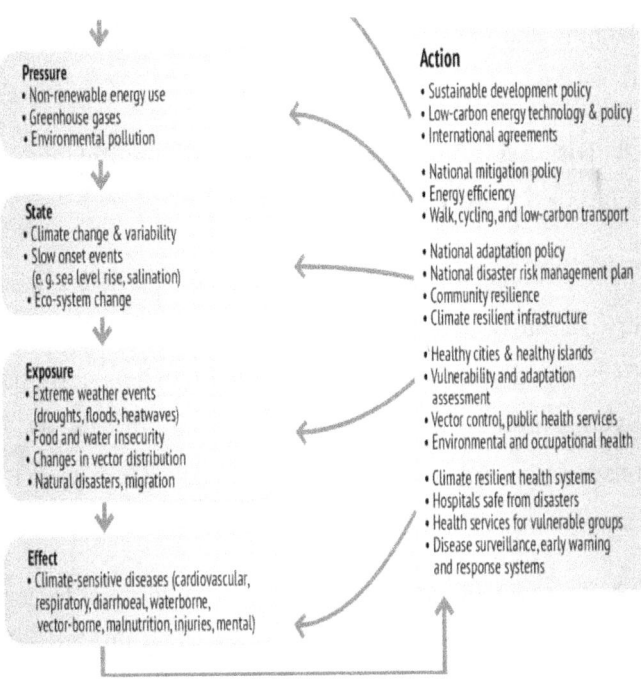

**Fig 5: The Driving Force-Pressure-State-Exposure-Effect-Action (DPSEEA) Framework for Climate Change**

The above DPSEEA Framework is widely implemented in EU (European Union) and other parts of the world by different industrial bodies to assess the climate change and its impact on human health.

That is why they have taken initiatives and urged other industries to do the same to change this orientation by making changes in different ETP and waste management planning and imposing climate change action policy (Post, 2015).

## 3.0 Discussion and Analysis of Its Significance in Relation to Government Action

The government has taken the issue of Climate change very seriously and they have acknowledged the harmful effect of industries on the climate. The Australian government has taken measures like limiting the water consumption for the industries and using ETP to treat the wastes thrown by the industries (CCES, 2019).

The government has also urged the industries to take initiatives to treat the harmful chemicals emitted by the exhaust of the industries as well. That is why the government can be credited to helping the climate for these courses of actions.

## 4.0 Forecast of Key Trends of the Climate Change and Its Relationship to the Organizational Policy Development

The climate change has made great impact on the world and has forced the industries to make certain changes in their development and action policy. The changes made for the company are given below:

- The company uses only the restricted water consumption level allowed for them.
- The waste products are no longer aimlessly thrown in the environment but treatment plan of those waste products is in action (Post, 2015).
- The harmful gases are not let to be emitted out in the environment rather the Effluent Treatment Plant (ETP) is used for that case (CCES, 2019).
- Measures are being taken so that the plastic bottles used are not harming the environment with the help of effective recycling.
- The heat level and increase of the heat are being kept under control.

# 5.0 Climate Change & Its Adverse Effects on Hobert City

In this section, we'll discuss the climate change and its adverse effect with a real-life scenario. To present this real-life scenario, we chose Australia's Tasmania capital Hobert City.

## 5.1 Natural Impact

### 5.1.1 Changes in Perception

The weather condition of the particular areas is greatly impacted by the climate change, it resulted change in perception to measure the temperature in the different seasons. In Hobart, the temperature of the summer is not similar like the past days. During 1960 the average summer season temperature was 11°C to 21°C but at present the average temperature is 13°C to 22°C (O'Gorman et al., 2016).

Some researches have already been identified that the current annual temperatures rise of the city is 0.4°C to 0.7°C. But in 2050 this annual temperature increasing rate can be 1 to 2°C, by 2100 the increasing rate can be touched 1 to 2.9°C land mark (Hobartcity, 2018). The consequence of this annual temperature rise can be inundated the 20% coastal areas of the city by 2100. Hot summer days and heat wave are more likely to happen in the upcoming years as the rise of temperature is continuing.

However, climate change has made the weather condition more extreme than ever before. During summer, extreme weather situations are seen in this city, sometimes the hot and humid weather situation is identified in this cold city (Hobartcity, 2018). Same type of weather condition is seen at the fall and spring seasons. This extreme weather condition is considered as a threat for the ecosystem and bio-diversity (Booth, 2016). Outbreaks of pest and any type contagious disease are imminent in this type of extreme weather condition.

## 5.1.2 Sea Level Rise

A recent research showed that the sea level of Tasmanian capital might be increased at 76cm by 2100 and inundated the coastal areas of the city. Thus, the coastal areas of the city such as Queens Domain, Glebe, New Town Bay etc. are at the coastal flooding risk. Other researches have suggested that the Hobart Airport, Royal Tasmanian Botanical Garden, AJ White Park, Maritime Museum of Tasmania etc. historical locations would be gone under water if the sea level rises by 1%.

**Fig 6: Predicted Inundation Scenario in Hobert by 2100**

We've predicted the inundation scenario in Hobert by 2100 with Costal Risk Simulation Application. Here, we identified the current days highest tide can submerge or flooded the coastal areas of the city. The situation could even worse if additional 0.74m tide is added with the current highest tide and this type of scenario can be seen by 2100 in Hobart.

## 5.2 Social Impact

### 5.2.1 Water Supplies

Climate change is resulting worst scenario in the different regions, the same type of situation might be seen in Hobart as well. Due to gradual sea level rise, salinity intrusion is more likely in this region (Marris, 2007). The increasing level of salinity can diminish the quality of pure water and create scarcity of pure drinking water for the living species.

For this reason, natural bio-diversity might be ruined, different species might be extinct due to this condition (Hughes, 2003). Soil moisture, fertility might be decreased due to salinity intrusion in the crop's lands. Incidence of forest can be substantially reduced due to the lack of fresh water in the region (McFarlane, 2014). The entire fresh water ecosystem is greatly affected with the scarcity of pure water.

### 5.2.2 Food Supplies

Soil moisture, fertility might be decreased due to salinity intrusion in the crop's lands, this may negatively affect the food supplies in the entire Tasmania region. Additionally, the rainfall pattern is gradually changing in the coastal and the central Tasmania regions (Mekanik & Imteaz, 2018). More rainfall has seen in the coastal regions, where less rainfall has seen in central Tasmania.

This extreme rainfall pattern is not good for agriculture, cultivation as more rainfall at the coastal zone can be flooded the area and less rainfall at the central part can be negatively affect the crops productions.

Tasmania's marine plants, natural fresh water reservoir and the animals are more likely to be affected due to extreme weather situation and lack of rainfall in the region (Mekanik & Imteaz, 2018). All of these situations might result food crisis in the entire regions.

### 5.2.3 Refugee Movement

Scarcity of fresh water and food may result refugee crisis in the Tasmania region, most of the communities may lose their habitat due to the climate change induced hazards and disasters effect. Coastal flooding and tidal surge would displace 50% people of Hobart City as 75% of the people of the city are lived at the coastal areas. They need to move to the other cities of Australia to find safe shelters (Birch, 2016).

The Intergovernmental Panel on Climate Change (IPCC) has suggested that 10 million people of the entire world can be lost their homes due to the climate change induced hazards and disasters effect by 2070. And the scenario might be seen in the upcoming years as the process has already begun. It will be tough for the Tasmania Govt. to safeguard all the people from the adverse effect of climate change.

## 6.0 Climate Change Adaptation Plan for Hobert City

### 6.1 Technological/ Structural

As 75% of the people of the Hobart city are living at the coastal zone, thus its necessary to relocate the buildings from the hazard zones (Palutikof, 2014). The govt. should help the community to identify the safest location in the coastal area.

In this context, tidal surge zones must be avoided and resilient building should be developed with sustainable building materials at the safest location from shoreline.

However, the infrastructures such as road, highway, bridge, hospital, school, cyclone shelter etc. should be resilient to coastal flooding, tidal surge and other climate change induced disasters.

## 6.2 Food & Water Management

Smart Agriculture practice can be taken by the Tasmania govt. across the region to introduce crops variety during the dry and pre-monsoon seasons. This agriculture practice ensures sustainable food supply at the moment of climate change (Tanaka, 2016). Adoption of different farming practices should be taken in the entire region to mitigate the adverse effect of climate change on the framing industry.

The water resources of Tasmania region are distributed unevenly, thus water shortage is imminent at the central part of Tasmania during less rainfall seasons (Palutikof, 2014). That's why the local govt. of Tasmania should implement a water management plan throughout the entire region.

They need to purify the sea water through water treatment plant to make it useable for agriculture, industrial use and framing. The natural water reservoirs must be monitored and maintained on regular basis to ensure free flow of water supply for the vulnerable communities.

## 6.3 Policy & Governance

The Tasmania govt. has already taken some steps to adopt the climate change, they are now providing climate change and impact of the change related information at regional and local levels (Hughes, 2003). Climate projection analysis data and the consequence of the climate change in the upcoming years data are now available for the public.

This effort might help the different industries and the vulnerable communities to adopt the climate change and developing new system to the mitigate the risk of the climate change. Tasmania Govt. is jointly working with the Australian central govt. to develop climate change action plan by considering risks and opportunities of the climate change (Hobartcity, 2018).

In this regard, the govt. is trying to assist the vulnerable communities to develop adaptative capacity and climate resilience across the region (Hobartcity, 2018). The govt. has already done some infrastructural research and analysis in the Hobart city to identify the vulnerable infrastructures at the coastal areas that can be damaged during climate change induced disasters.

## 7.0 Conclusion

Climate change is also causing global warming across the different regions, all of these are happening in real time. This change is resulting sea level rise, extreme weather condition, excessive humidity in the air during summer etc. in the effected regions.

The people of different urban and rural areas are suffering much due to the adverse effect of climate change. Because frequency, intensity and the severity of any hazard or disaster is enhancing as a result of climate change.

Finally, it can be said that the effect of the company on the environment is very harmful but if the right measures are taken to solve or minimize the effect of these issues then a big change can be made. The different industries that the company belonged can help to make a big change to save the environment as well.

Meanwhile, the Tasmanian govt. should implement their climate change action plan/ strategy immediately to safeguard the vulnerable communities from the adverse effect of climate change.

## 8.0 References

ANU. (2019). - Global Warming and Climate Change: what Australia knew and buried - ANU. Retrieved from https://press-files.anu.edu.au/downloads/press/p303951/html/Chapt07.xhtml?referer=&page=12

Australia and New Zealand, 1800–1945. *Wires Climate Change*, *7*(6), 893-909. https://doi.org/10.1002/wcc.426

Birch, T. (2016). Climate Change, Mining and Traditional Indigenous Knowledge in Australia. *Social Inclusion*, *4*(1), 92-101. https://doi.org/10.17645/si.v4i1.442

Booth, T. (2012). Biodiversity and Climate Change Adaptation in Australia: Strategy and Research Developments. *Advances In Climate Change Research*, *3*(1), 12-21. https://doi.org/10.3724/sp.j.1248.2012.00012

Crowley, K. (2017). Up and down with climate politics 2013–2016: the repeal of carbon pricing in

Australia. *Wires Climate Change*, *8*(3). https://doi.org/10.1002/wcc.458

Cave, D. (2019). It Was Supposed to Be Australia's Climate Change Election. What Happened?. Retrieved from https://www.nytimes.com/2019/05/19/world/australia/election-climate-change.html

CCES. (2019). Business Strategies to Address Climate Change | Center for Climate and Energy Solutions. Retrieved from https://www.c2es.org/content/business-strategies-to-address-climate-change/

Cha, J., & Lee, H. (2017). The Impact of Climate Change Awareness on Demand for Climate Change Response. *Journal Of Environmental Policy And Administration*, *25*(4), 63-77. doi: 10.15301/jepa.2017.25.4.63

Droege, S. (2012). The challenge of reconciliation: climate change, development, and international trade. *Climate Policy*, *12*(4), 524-526. doi: 10.1080/14693062.2012.688523

Hobartcity. (2018). *Sustainable Hobart Action Plan*. Hobartcity.com.au. Retrieved 15 July 2021, from https://www.hobartcity.com.au/Council/Strategies-and-plans/Hobarts-Climate-Change-Strategy.

Hughes, L. (2003). Climate change and Australia: Trends, projections and impacts. *Austral Ecology*, *28*(4), 423-443. https://doi.org/10.1046/j.1442-9993.2003.01300.x

Lemos, M., & Rood, R. (2010). Climate projections and their impact on policy and practice. *Wiley Interdisciplinary Reviews: Climate Change*, *1*(5), 670-682. doi: 10.1002/wcc.71

Marris, E. (2007). Australia warms to climate change. *Nature Climate Change*, *1*(711), 90-90. https://doi.org/10.1038/climate.2007.62

McFarlane, D. (2012). Projected climate change impacts on water resources in south-western Australia. *Journal Of Earth Science & Climatic Change*, *01*(S1). https://doi.org/10.4172/2157-7617.s1.004

Mekanik, F., & Imteaz, M. (2018). Variability of cool seasonal rainfall associated with Indo-Pacific climate modes: case study of Victoria, Australia. *Journal Of Water And Climate Change*, *9*(3), 584-597. https://doi.org/10.2166/wcc.2018.146

NASA. (2021). *Deadly Floods Surprise Europe*. Earthobservatory.nasa.gov. Retrieved 25 July 2021, from https://earthobservatory.nasa.gov/images/148598/deadly-floods-surprise-europe.

Oxfam. (2021). *5 natural disasters that beg for climate action | Oxfam International*. Oxfam International. Retrieved 25 July 2021, from https://www.oxfam.org/en/5-natural-disasters-beg-climate-action.

O'Gorman, E., Beattie, J., & Henry, M. (2016). Histories of climate, science, and colonization in

Palutikof, J. (2010). The view from the front line: Adapting Australia to climate change. *Global Environmental Change*, *20*(2), 218-219. https://doi.org/10.1016/j.gloenvcha.2010.03.002

Post, M. (2015). Multi-organizational Alliances and Policy Change: Understanding the Mobilization and

Impact of Grassroots Coalitions. *Nonprofit Policy Forum*, *6*(3). doi: 10.1515/npf-2014-0030

Slezak, M. (2019). David Attenborough has a message for our leaders. Did they hear it?. Retrieved from https://www.abc.net.au/news/2019-05-07/climate-change-federal-election-morrison-shorten-policies-votes/11084580

WTO. (2019). WTO | Trade and environment - WTO and the challenge of climate change. Retrieved from https://www.wto.org/english/tratop_e/envir_e/climate_challenge_e.htm

# About the Authors

## G. Hossain

Ghazi Mokammel Hossain is a freelance writer. He has written some books as well as articles, research papers and creative articles. The author is also working in different type of researcher projects. He was born on 31 December, 1993. He has passed his S.S.C exam in 2008 and passed H.S.C exam in 2010. He has graduated with a Bachelor's of Business Administration (BBA) in HRM in 2015 from a renowned University.

He has also completed Social Compliance & CSR diploma in 2016. From Dhaka University he has completed Masters in Disaster Management (MDM). He published his first book called "IPv4 IP6 Technology & Implementation" in Amazon Kindle and Createspace on 201. After that, he has already been published different books on various subjects.

The author published an outstanding thrilling novel called "Anwar: Emergence of Unknown Defenders" in 2016 on Amazon kindle and Createspace.

Playing football, Cricket, PC games, reading books, novels, research papers, cycling and mountain climbing are his favorite hobbies.

## Mohammed Fazle Mubin

Mohammed Fazle Mubin is an editor, research, article and creative writer. He has contributed in writing many articles, research papers, books and analysis papers.

He has passed SSC from Motijheel Ideal School and his HSC from Notre Dame College both under Dhaka board. He is currently doing BSc on Computer Science & Engineering in a well renowned University of Bangladesh.

# Also by G. Hossain & GM Publishers

(Including Purchase Link QR Code)

Notebook: Colored Retro Styled Blank Line Pocket Notebook Paperback – July 25, 2021 by G. Hossain

Lightweight Cryptography & Crypto Currency: Possibilities & Challenges in the Modern Business Context - July 12, 2021 by G. Hossain, Sadman Alam, Md. Fazle Mubin

The Author: Revealer of Unseen Truth (Day to Dusk Book 1)- December 17, 2019 by G. Hossain

Quick & Easy Pickling Cookbook: With Chutney, Jelly & Sauce Recipes- December 11, 2018 by S.T. Ara

True Friends- July 20, 2018 by G. Hossain & MD. Fazle Mubin

Marketing Strategy & Research: In the Context of Different Organizations- November, 2017 by Ghazi Mokammel Hossain & MD. Fazle Mubin

Supermarket Management Practices: In the Changing Economic Environment- November, 2016 by Ghazi Mokammel Hossain

Anwar: Emergence of Unknown Defenders- August 10, 2016 by Ghazi Mokammel Hossain

The Survival of USA – Part Two: A Novel - August, 2016 by Ghazi Mokammel Hossain & MD. Fazle Mubin

Business Environment: Theoretical & Organizational Aspects – July,2016 by Ghazi Mokammel Hossain

The Survival of USA - Part One: A Novel – March, 2016 by Ghazi Mokammel Hossain, MD. Fazle Mubin & Pranjal Rahman

Enterprise IPv6 for Enterprise Networks- December, 2015 by Ghazi Mokammel Hossain & Fathe Mubin

Heart of Democracy: A Versatile Poetry Book - Aug 28, 2015 by Ghazi Mozammel Hossain

The Brave Parrot of Jungle - Dec 11, 2014 by S.T. Ara & Gulshan Ahmed

The Mirror of Religion - Jul 19, 2015 by Ghazi Mozammel Hossain & Richard Marks

Introduction to Network on Chip Routing Algorithms - Oct 4, 2014 by Ghazi Mokammel Hossain

Ebola Epidemic: A Detail Survival Guide From Ebola Virus Disease Outbreak - Oct 25, 2014 by Ghazi Mokammel Hossain
Fundamental of API Based Financial Engineering - Oct 17, 2014 by Ghazi Mokammel Hossain

IPv4 IPv6 Technology and Implementation - Nov 2, 2013 by Ghazi Mokammel Hossain & GM Hossain

For more information, please visit Amazon Author Central
**GM Publishers**
https://www.amazon.com/Ghazi-Mokammel-Hossain/e/B00GGATR2K

www.ingramcontent.com/pod-product-compliance
Lightning Source LLC
Chambersburg PA
CBHW070843220526
45466CB00002B/876